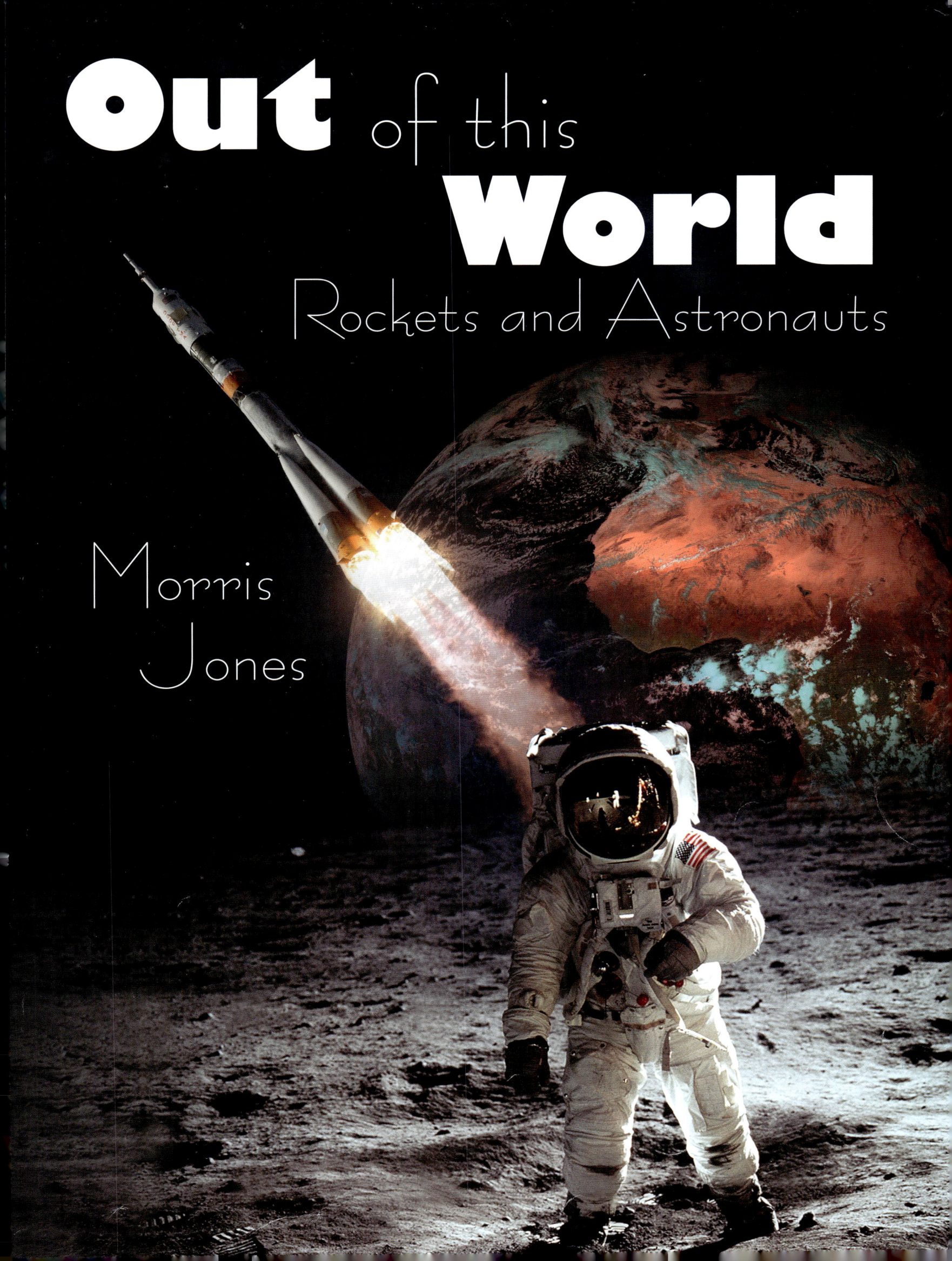

Out of this World

Rockets and Astronauts

Morris Jones

First published in Australia in 2008 by
Young Reed an imprint of New Holland Publishers (Australia) Pty Ltd
Sydney • Auckland • London • Cape Town

1/66 Gibbes St Chatswood NSW 2067 Australia
218 Lake Road Northcote Auckland 0627 New Zealand
86 Edgware Road London W2 2EA United Kingdom
80 McKenzie Street Cape Town 8001 South Africa

10 9 8 7 6 5 4 3 2 1

National Library of Australia Cataloguing-in-Publication entry

Author: Jones, Morris, 1971-

Title: Out of this world : rockets and astronauts
                     / Morris Jones.
ISBN: 9781921073397 (hbk.)
Notes: Includes index.
                     Bibliography.
Target Audience: For primary age children.
Subjects: Astronauts.
                     Manned space flight.
                     Astronautics.
Dewey Number: 629.45

Commissioning Editor: Yani Silvana
Designer: Hayley Norman
Production: Linda Bottari
Printer: Tien Wah Press, Malaysia

Picture Credits

Contents

What is Space?

Space is sometimes called 'outer space' because it is the region outside the Earth's **atmosphere**. Space is a huge area filled with almost nothing. There are very few objects and no air. For this reason, space is sometimes called a 'vacuum'. You could not live in space without special equipment to protect you. A supply of oxygen to breathe and protection from the extreme temperatures are two things you would need. There are no other planets that we know of where a human could live without this type of equipment. Space is dangerous.

Where Does Space Start?

As you travel away from the Earth, the air gradually gets thinner. A few kilometres above the ground, it is so thin that it is difficult to breathe. The higher you go, the thinner the air becomes. By the time you are about 100 kilometres above the ground, there is almost no air around you. Scientists generally agree that space starts at this distance from the Earth.

What Does Earth Look Like from Space?

If you travelled into space, the sky around you would look black. That is because there is no air to scatter the sunlight. On Earth, we normally have blue skies because the light is scattered by molecules in the air. Looking back at the Earth, you would see that it is round and looks blue, because most of it is covered with water. You would also see clouds. Earth is like a big ball travelling through space as it **orbits** the Sun. Other planets also go round the Sun, but the distances between them are huge. Close to the Earth they just look like small dots in the sky. You can sometimes see stars, but they don't look much different from the planets.

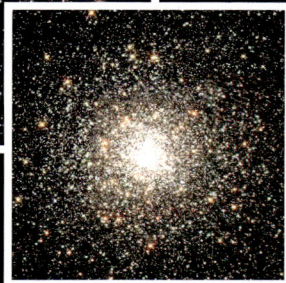

What Else Is Out There?

Space is almost empty but it is still filled with things that are difficult to see, like energy. As this invisible energy travels through space it can heat up your spacecraft. There are also tiny particles, smaller than atoms, that travel rapidly. The energy and the particles are sometimes known as 'radiation'. Some of this radiation can harm you if you are exposed to too much of it.

Other Suns

The stars you can see in the night sky are actually like our own Sun. Some of these stars are hundreds of times larger than the Sun but they are so far away that they don't light up the Earth. Scientists know that there are other planets orbiting these stars. There are other **solar systems** around other stars. These stars and their planets are so distant that no spacecraft has ever gone there. We can only look at them with powerful telescopes.

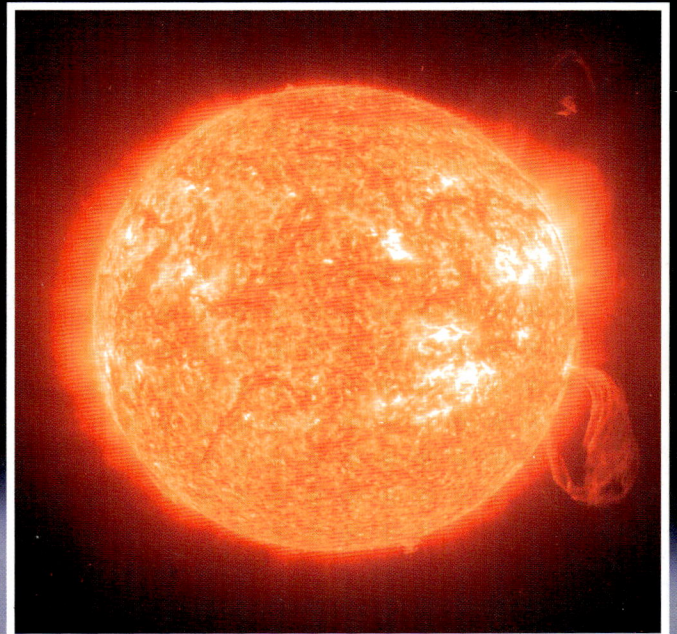

Why is the Sky Blue?

On a clear day, light from the Sun is scattered by atoms in the atmosphere. Red, yellow and green light is not scattered as well as blue light—so the sky looks blue. Towards evening, however, the red and yellow light is scattered more, so we get beautiful sunsets.

What Drives a Rocket?

Most rockets use liquid fuels. This gives the rocket a lot of power. They also carry oxygen in liquid form in a tank so the fuel can burn. The oxygen has been cooled so much that it changes from a gas to a liquid. That way more oxygen can be carried because a liquid takes up less room than a gas. Instead of oxygen some rockets use other chemicals to burn their fuel. These may be easier to load into the rocket than liquid oxygen, which has to be kept very cold, but they do not produce as much power as liquid oxygen.

Why Do Rockets Need So Much Fuel?

Rockets need to travel very fast in order to reach outer space. A satellite in **orbit** needs to travel more than 8 kilometres a second to stay in space. That's many times the speed of sound. No aircraft can fly that fast. If a rocket is carrying something beyond the Earth, it needs to travel even faster— more than 11 kilometres per second. That is how fast you need to go if you want to escape the Earth's gravity to reach the Moon or another planet.

Solid Rockets

The simplest rocket does not contain fuel tanks for liquid fuel. It is a tube lined with solid fuel. Solid rocket fuel is like explosive, and it produces a lot of force. It can be a mixture of chemicals like those you'd find in gunpowder. But solid-fuelled rockets are not as powerful as liquid-fuelled rockets.

What's with the Pointy End?

So it can move through the air smoothly, the top of the rocket is shaped like a cone. This is really a shell that covers the satellite or spacecraft that is on top of the rocket. The shell is known as the 'payload shroud'. When the rocket reaches space, the payload shroud falls away.

Rockets

A rocket is a vehicle that can take people and objects into space. Most rockets are like long tubes filled with fuel with at least one rocket engine at their base. They don't normally have much room on board for satellites, **astronauts** or other cargo. The fuel takes up most of the space inside the rocket.

We Have Lift Off!

Most rockets used to reach space are like several rockets joined together. These are known as 'stages' in a rocket. The stages are usually stacked on top of each other. When the rocket takes off, the first 'stage' fires until it uses up all its fuel. Then it falls away from the rocket and the second 'stage' starts to fire, sending the rocket higher. Using stages saves weight because the used stages don't need to be carried all the way into space. This means that more cargo, or 'payload', can be carried.

Rocket Boosters

Some rockets have smaller rockets strapped to their sides. These fire at the same time as the first stage to give the rocket more power. They are known as 'boosters'. Some rockets have so many small boosters that they look like bundles of pipes.

Russians Head the Race

Yuri Gagarin was the first human to travel in space. He was launched into space on board *Vostok 1* on 12 April 1961. He made one **orbit** of the Earth and returned.

First Space Walk

The first object to land safely on the Moon was a robot spacecraft called Luna 9. It touched down on 3 February 1966, and transmitted pictures from the surface of the Moon. It was another first for Russia.

First Space Walk

On 18 March 1965 Alexei Leonov became the first person to step out of a spacecraft and walk in space. He was a Russian **cosmonaut**, which is what the Russians call their **astronauts**, and he wore a spacesuit that allowed him to breathe oxygen while he floated outside his spacecraft, *Voskhod 2*.

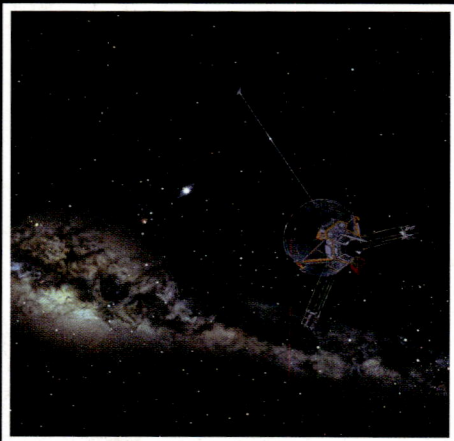

First Object to Go Beyond the Planets

The first spacecraft to go beyond the planets of the **solar system** was the *Pioneer 10* robot space probe. Launched by the USA, *Pioneer 10* flew past Jupiter in 1973, and kept travelling deeper into space. It crossed the orbit of Neptune, the most distant planet in the solar system, on 13 June 1983. *Pioneer 10* no longer functions, but it is still travelling out into deep space.

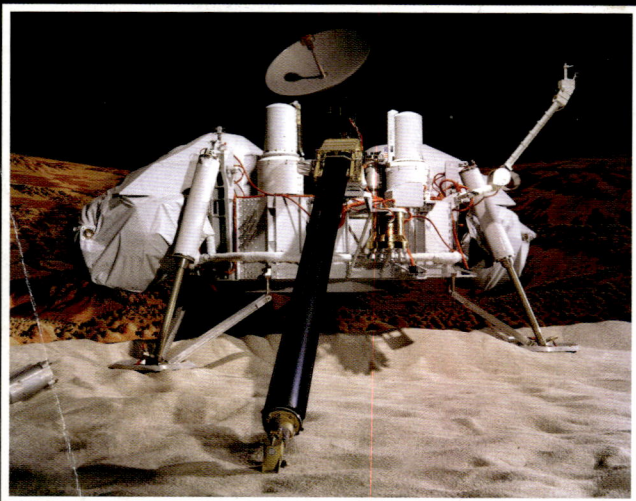

First Landing on Mars

The Americans had another first with the Viking 1 Lander, the first object to successfully land and operate on the surface of Mars. It touched down on the Martian surface on 20 July 1976. It sent back pictures of the red sands and rocks of Mars, and scooped up some soil with a robot arm. There were no **astronauts** on board.

First in Space

The first satellite to leave the Earth and travel in space was *Sputnik*. The Russians launched it on 4 October 1957. *Sputnik* was a simple metal sphere with a radio transmitter inside. It had four antennas and gave out simple 'beep' signals. *Sputnik* amazed the world because it proved that space travel was more than just a dream.

Space Dog

The first living creature to go into space was a dog called Laika. She was placed on board *Sputnik 2*, the world's second satellite, which was launched on 3 November 1957. Sadly, Laika died after just a few hours in space when her life support system failed.

SHQIPERIA   3 NËNDOR 1957
LAIKA   1

First Astronauts on the Moon

Two Americans, Neil Armstrong and Buzz Aldrin, were the first **astronauts** to set foot on the Moon. They touched down on 20 July 1969. They were part of the *Apollo 11* mission. You can find out more about this mission in the section 'First Landing on the Moon' on page 16.

Let's Go to Mars

Mars is very far away and it would take a long time to get there. So far no humans have tried to travel to Mars or any other planet. Nobody knows who those first explorers will be, or when they will go. It will be many years before anyone is sent there. Perhaps you could be one of them.

Satellites

A satellite is an object that moves in **orbit** around something else. The Moon is a satellite of the Earth. The Earth is a satellite of the Sun. But when people talk about satellites, they usually mean 'artificial satellites'. These are spacecraft that have been placed in orbit around the Earth.

Weather Watch

Satellites sometimes look back at the Earth and see what's happening. They watch clouds and weather patterns and they can see dangerous weather events such as tornadoes and hurricanes. These are huge storms that look like swirling circles of cloud. Seen from space they look beautiful, but they can destroy entire cities. When a satellite photograph shows a powerful storm, emergency services can prepare for its arrival, before it damages anything.

What's an Orbit?

When an object that is travelling very fast is held by the gravity of a planet, it moves in a path that bends around the planet. Sometimes this looks like a circle, but **orbits** can also be egg-shaped.

Watching the Stars

Some satellites carry telescopes to look at other objects in space. The telescopes work very well in space because they are outside the Earth's **atmosphere**. There is no pollution or interference from air molecules to block their view. These telescopes can sometimes see stars and other objects that can't be seen from telescopes on Earth.

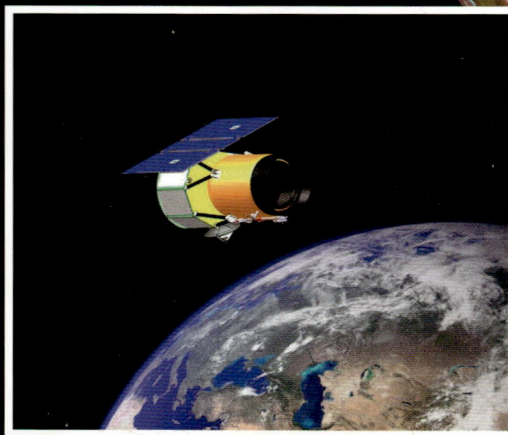

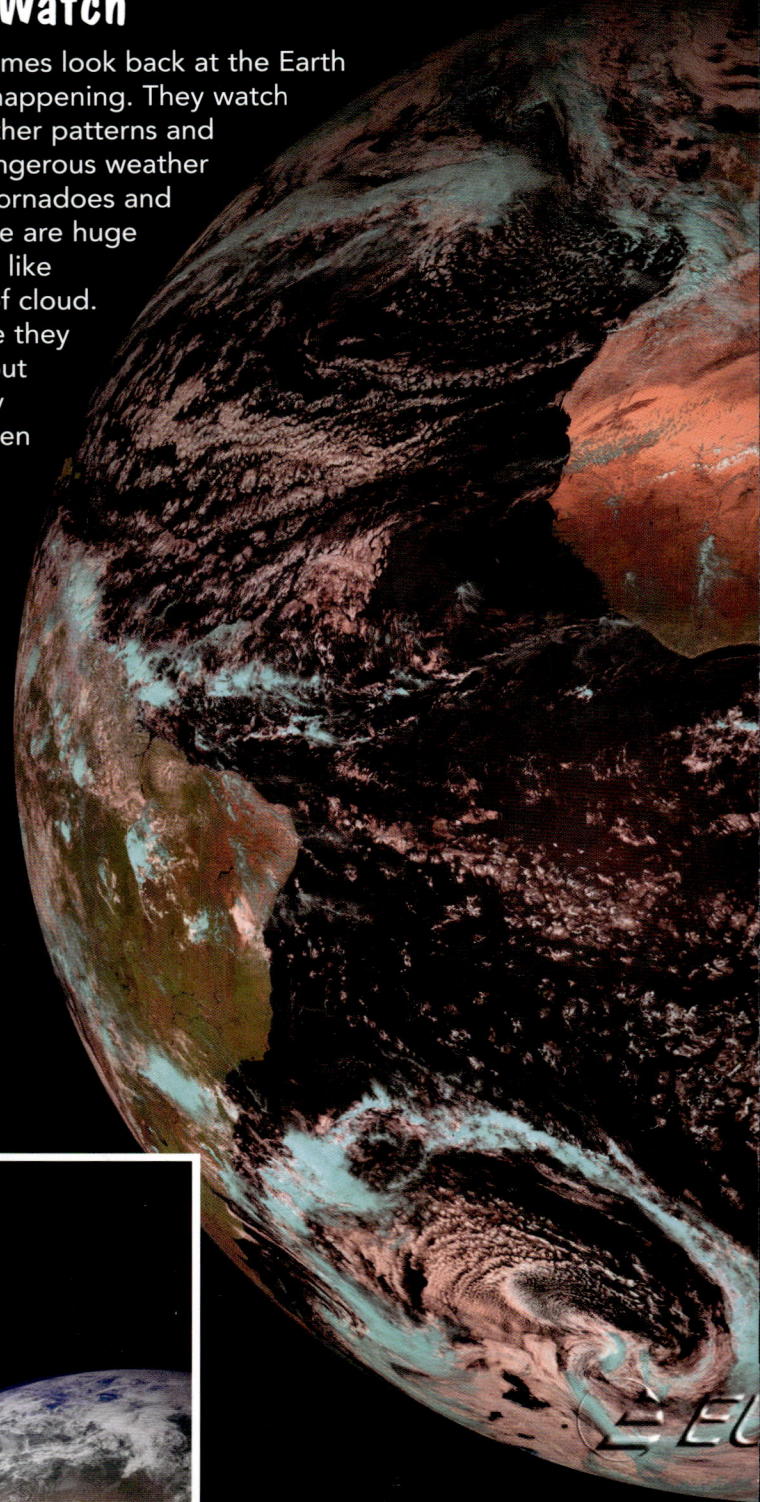

What Happens to all that Junk?

There is a lot of junk floating about in space. Most satellites never return to the ground after they are launched. Sometimes, when they are no longer needed, they are steered back into the **atmosphere**. Because they are travelling so fast they 'burn up' in the air and are destroyed. But sometimes bits of large satellites don't burn up entirely and crash into the Earth.

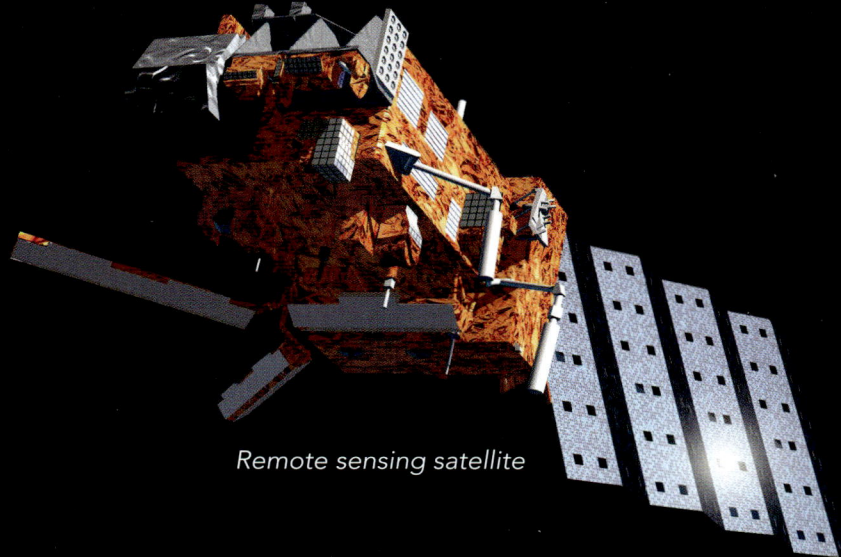

Remote sensing satellite

Keeping in Touch

A communications satellite is like a big antenna in the sky. It can pick up radio signals and send them from one place on Earth to another. This is how a lot of television is transmitted. Communications satellites also carry some telephone calls and Internet downloads. Look around and you're sure to see satellite dishes or antennas on a roof near where you live.

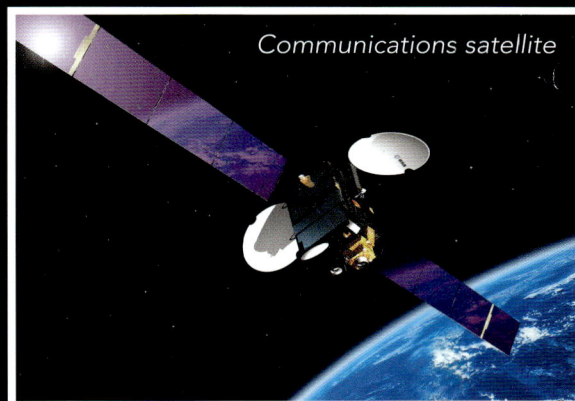

Communications satellite

Sssshhh ... Eye in the Sky

Some satellites have powerful cameras that can focus in on buildings and equipment. They are used to track navy ships, tanks and other military equipment. This lets countries know if another country is secretly preparing to fight a war. These satellites are known as 'spy satellites' and are usually kept secret. Governments do not usually talk about having these satellites or what they are seeing.

Using the Sun's Energy

Most satellites have solar panels. These are flat panels, covered with special materials that turn sunlight into electricity. This is how the satellites generate their power. The electricity runs the equipment on board and allows the satellites to send radio transmissions back to Earth.

11

The Space Shuttle

For many years, the only way to get into space was with an ordinary rocket. Each rocket could only be used for one flight. At the end of its mission, it was usually destroyed in the Earth's **atmosphere**, or crashed back to the ground. You can imagine how expensive it would be to fly anywhere if you could only use an aircraft for one trip. Space travel was super-expensive, because every new journey required a new rocket.

A Re-useable Spacecraft

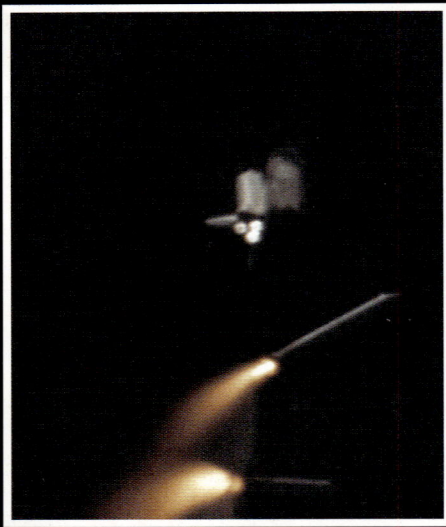

Scientists proposed building a space shuttle as a way of making space travel easier. The space shuttle looks like a jet aircraft, with triangular wings. It is strapped to the side of a huge fuel tank, and two rocket boosters are added to the side of the tank. The rocket motors allow the space shuttle to take off. As the space shuttle rises higher into the **atmosphere**, its rocket boosters separate and fall into the ocean on parachutes. Ships recover them from the ocean so they can be re-used. Soon afterwards, the huge fuel tank separates and is not recovered. It falls into the atmosphere and breaks up.

What Happens Next?

The space shuttle then goes into **orbit** around the Earth. When it is time for the mission to end, the space shuttle re-enters the atmosphere. It glides like an aircraft to a runway, and lands. The space shuttle, and the rocket boosters that helped to launch it, can be used again for another space mission. Another huge fuel tank must be built for the next launch, but the rest of the system can fly many times. By re-using the space shuttle and its rocket boosters, it was hoped that space travel would become easier and cheaper.

Many Jobs

The space shuttle can carry as many as eight **astronauts**. It also has a huge cargo bay, for taking other objects into space. The space shuttle has been used to carry satellites into orbit, and to transport parts and supplies for the International Space Station (see pages 12–13). It can also bring large objects back from space. Astronauts have spacewalked from the space shuttle to carry out repairs on satellites.

Problems

The space shuttle was first launched in 1981 so it has flown for a long time. However, it does not operate as well as its designers had hoped. It is very expensive to repair between launches and it is also complicated to fly. This means it has not succeeded in making space travel cheaper and easier. Two space shuttles have also been destroyed in flight. In 1986, the space shuttle *Challenger* exploded during launch, killing all seven **astronauts** on board. In 2003, the space shuttle *Columbia* was destroyed as it re-entered the **atmosphere** at the end of its mission. Another seven astronauts were killed. There are currently three space shuttles in operation: *Discovery*, *Atlantis* and *Endeavour*. The National Aeronautics and Space Administration (NASA) the American space agency that operates the space shuttle, has decided to stop launching the shuttle in a few years. The space shuttle will probably make its last flight by 2011.

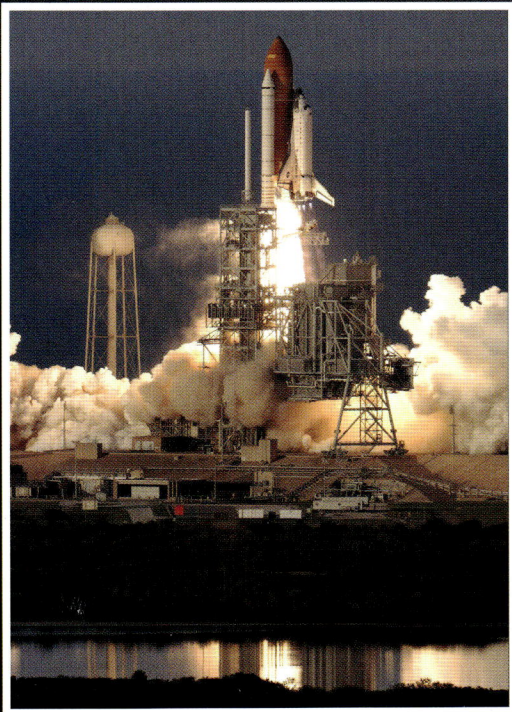

A New Spacecraft

NASA is developing a new spacecraft to replace the space shuttle. It is sometimes called the Crew Exploration Vehicle, but its official name is *Orion*. *Orion* is a small, cone-shaped capsule, designed to be launched on top of a rocket. It does not have wings, but lands with parachutes. It will be reusable. *Orion* should start carrying astronauts into space around the year 2014.

The International Space Station

The only space station operating right now is the International Space Station. It has been in **orbit** since 1998, and will continue to work for many years. It's the largest thing orbiting the Earth, apart from the Moon. Many nations, including America, Russia, France, Japan and Canada, have helped to build it, so they can all send astronauts there. Parts for the space station have usually been carried into space by the space shuttle. Russian rockets have also transported some parts. Astronauts have joined the parts together to make a large station that is too big to be launched by a single rocket.

Getting There

You travel to the International Space Station on board a space shuttle, which flies up to the space station and docks with it. When you get there you go through a tunnel that connects the cabin of the space shuttle to the space station. You can just float through it to enter the station. When you arrive, other astronauts already on the station are preparing to come home. One of them takes your seat on the space shuttle. When it separates from the space station and returns to Earth, they go and you stay.

Space Stations

A space station is like a house in space. It allows astronauts to live in space for several months. You would be more comfortable on board a space station than other types of spacecraft such as the space shuttle. There is more room, and you would have a bathroom, a small bedroom and a small living area. It needs to be more comfortable because astronauts have to stay aboard for so long.

Building the Station

The space station is gradually growing as space shuttles bring more parts into space. Sometimes, the shuttle will add another living module for the crew. Sometimes it carries more solar panels for generating electricity. Often, the space shuttle just delivers more food and equipment inside a large cylindrical cargo module. Other small spacecraft from Russia, Europe and Japan also carry cargo to the space station.

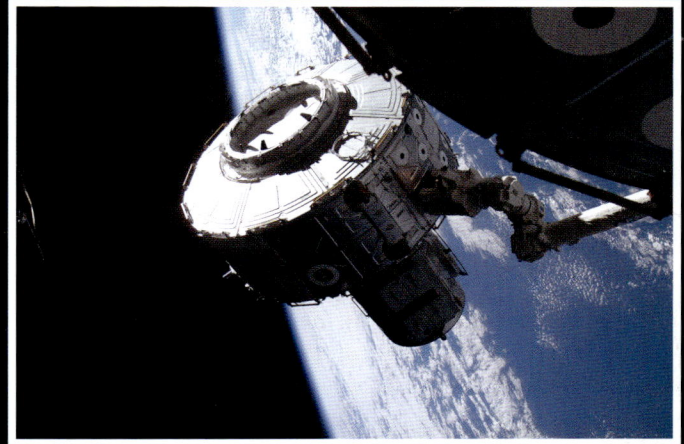

Living There

The crew sections of the space station are like huge metal pipes, joined together like rooms in a house. Some of them are used for eating and sleeping. Other sections contain the tools and equipment you need to work with. There are many parts on the space station, and something usually needs to be repaired, so some of your time would be spent fixing the equipment when it breaks down.

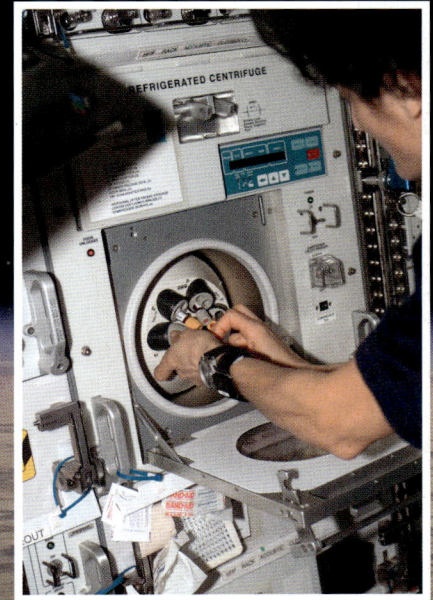

Coming Home

Sometimes, there is no space shuttle ready to fly when the space station crew needs to be changed. In that case, even though you may have flown up on the shuttle, you would have to come back to Earth on board a Russian Soyuz capsule spacecraft. There is always one of these docked to the space station, in case the crew need to return to Earth quickly.

The Apollo II Lunar Module prepares to land.

The Apollo Program

The program that would send astronauts to the Moon and bring them back was called Apollo. It produced a giant rocket, the Saturn 5, to send them there. This was the largest and most powerful rocket to be successfully operated over a long series of flights. It was tall, standing more than 100 metres high. That's taller than most buildings! To get to the top of the rocket astronauts had to ride up in an elevator.

Apollo 11

The first *Apollo* missions had tested the Apollo spacecraft to see if a landing on the Moon was possible. The first Apollo mission to attempt to land astronauts was *Apollo 11*. Neil Armstrong, Buzz Aldrin and Michael Collins were the astronauts chosen for the mission. All were trained jet fighter pilots with engineering degrees and they had all flown in space before.

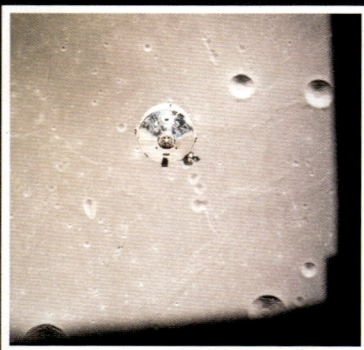

Command Module, Lunar Module

The Apollo program also had two spaceships that were sent to the Moon. One was called the command module. It would carry three astronauts to the Moon, then go into **orbit** around the Moon. The second spacecraft was called the lunar module. It would take two of the astronauts down to land on the Moon and return them to the command module.

The Landing

Once the command module, *Columbia*, was in orbit around the Moon, Neil Armstrong and Buzz Aldrin entered the lunar module, *Eagle*. Then they flew down to the surface of the Moon and, with less than 30 seconds fuel left, the Eagle landed. Six hours later, on 20 July 1969, Neil Armstrong became the first person to step onto the Moon. Buzz Aldrin also walked on the Moon. The astronauts set up experiments on the surface and collected rocks and soil samples. After spending about two and a half hours on the lunar surface they climbed back into the lunar module.

First Landing
on the Moon

When the space age began, some people thought that astronauts would not be able to land on the Moon. They thought that any spacecraft that landed there would sink into the dust. But America was determined to send astronauts there. They knew that Russia had achieved most of the 'firsts' in space including the first satellite, the first living creature and the first human in space. America was behind in the space race and wanted to catch up.

Home Again

The astronauts had a few hours' sleep in the lunar module, then they took off from the Moon. Michael Collins was waiting for them in the command module, which was orbiting the Moon. Armstrong and Aldrin moved back into the command module, and the lunar module was released to crash into the Moon. Then the command module fired its engines to break out of its orbit round the Moon and headed back to Earth. It splashed down safely in the Pacific Ocean southwest of Hawaii. The journey had taken just over 18 days and covered 1 764 400 kilometres. They had made history.

The Later Apollo Missions

The flight of *Apollo 11* showed that astronauts could explore the Moon, so more Apollo missions were sent there. These later missions performed more tasks on the Moon than *Apollo 11* and carried more equipment.

Lunar Rover

Apollo 15, Apollo 16 and *Apollo 17* took a small car to the Moon. This was called the Lunar Rover. It was a very simple vehicle with no doors or roof. The two astronauts sat in the open, in their spacesuits. The Lunar Rover was powered by electric motors and batteries. It allowed the astronauts to travel further than they could walk.

Lunar Experiments

The astronauts set up some scientific equipment on the Moon. The experiments kept working after the astronauts left the Moon. Some measured 'moonquakes' and the temperature of the ground. The information they collected was transmitted back to Earth by radio.

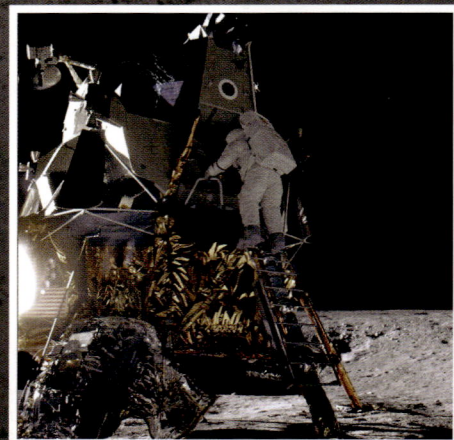

Six Landings—Twelve Astronauts

Apollo 12 was the next mission to land on the Moon in 1969. *Apollo 13* had problems and returned to Earth without attempting a lunar landing. *Apollo 14*, *Apollo 15*, *Apollo 16* and *Apollo 17* all landed astronauts on the Moon. The later Apollo missions used the same type of Saturn 5 rocket that had launched *Apollo 11*. They also used the same type of command module and lunar module spacecraft. But some minor improvements were made for each mission. After 1972, there were no more Apollo missions to the Moon although three more had been planned. The six Apollo landings on the Moon took twelve astronauts to the surface of our nearest neighbour in space. Since then, nobody else has walked on the Moon.

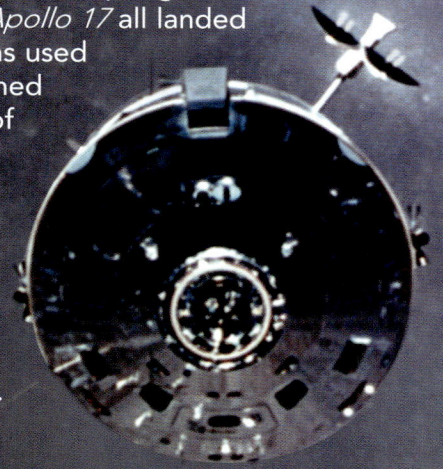

Sleepovers on the Moon

The first astronauts to walk on the Moon on *Apollo 11* did not stay on the surface very long. But the final Apollo missions stayed about three days. The astronauts spent hours walking and driving on the Moon. Then they went back to the lunar module to eat and sleep before going out again the next day.

Moon Rocks

More than 380 kilograms of rocks were brought back to Earth by the six Apollo missions. Scientists have found that some of the Moon rocks were formed by volcanoes, just like rocks on Earth. Because there is no air on the Moon the rocks have to be kept in special sealed chambers that do not allow them to come in contact with oxygen. This keeps the rocks in their original state because contact with oxygen would change their chemical structure. Moon rocks have also been sent to museums around the world. Find out if there are any in your local museum.

Lunar Lasers

One experiment was a panel of laser reflectors, which work like mirrors. A laser beam was fired from Earth at the reflector. It bounced off the reflector and came back to Earth. By measuring the time it took the laser beam to travel to the Moon and back, scientists could work out the exact distance to the Moon. The laser reflectors are still being used.

Mars by Machine

It's much easier to explore Mars with machines. Robot space probes do not need food, oxygen or life-support systems so they work much better on long space missions. But Mars is dangerous even for robot space probes. Many missions launched to Mars have failed to work properly.

It's a Long Way Away

Mars is further from the Sun than Earth. It takes months to send a spacecraft to Mars, and it would take more than two years to make the round trip from Earth to Mars and back again. Keeping astronauts alive and well during such a long journey into space would be difficult. That is why no astronauts have been sent there yet.

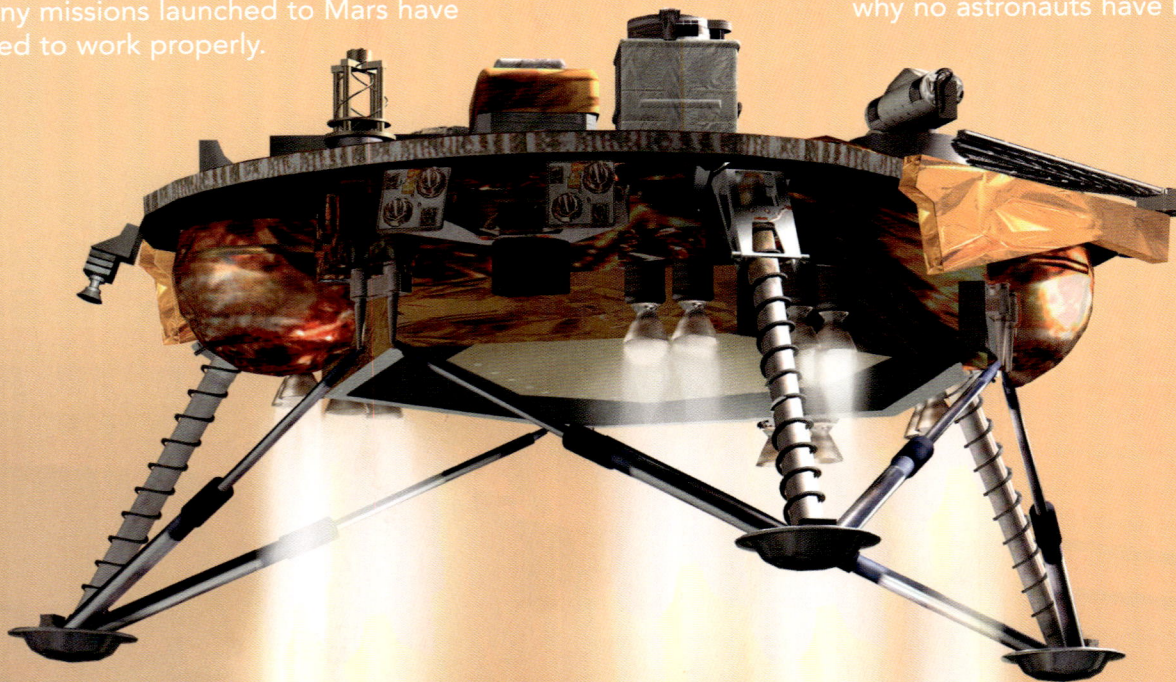

Rovers on Mars

The best way to explore the surface of Mars is with a **rover**. This is a robot vehicle with wheels that can drive across huge distances, and see many different areas. They are 'driven' by an operator from a control room on Earth. Cameras on board the rover send images back so you can see where it is. You can then send driving instructions to the rover via a radio link telling it which areas in the photos to visit next. This is a slow process. It takes several minutes for a radio signal to travel to Mars, and the rovers don't drive very fast.

Bigger Rovers

The most famous rovers on Mars are the two Mars Exploration Rovers. These vehicles, Spirit and Opportunity, landed on Mars in 2004 and have driven across Mars for years. Larger rovers will explore Mars in the future, such as the Mars Science Laboratory due to be launched in 2009.

Martian Farmers

Back in the 19th and early 20th century people thought there were canals on Mars. They thought perhaps they were irrigation canals carrying water to Martian farms. The arrival of the *Mariner 4* space probe in 1965 proved there were no Martian farmers or canals. It sent back pictures of a barren, waterless landscape.

Exploring Mars

Mars is a small, rocky planet that is covered with red dust. It looks like a desert and there are no lakes or oceans. The sky is usually pink, because it is filled with fine dust. Mars is a very cold planet and its air is very thin. There is no oxygen in its **atmosphere**, most of which is carbon dioxide.

What Can You See?

A spacecraft orbiting Mars can look down on the planet and transmit images back to Earth. You can see huge volcanoes on its surface. These are no longer active, but they look impressive. You can also see a huge canyon that's as wide as Australia. Sometimes you wouldn't see anything at all because huge dust storms cover the planet, blocking everything from view.

Martian Ice

If you landed on Mars and dug into the ground, you would find huge amounts of ice just a few metres below the surface. There's one problem—it's filthy. Mixed in with the ice is dust and rock so it looks like frozen mud. There are also ice caps at the north and south poles of Mars, just like we have on Earth. The ice here looks much cleaner. Some of it is made of frozen water, and some of it is carbon dioxide that is so cold it has frozen solid.

Visiting Jupiter and Saturn

Jupiter and Saturn are huge planets that are made mostly of gas so they are sometimes called 'gas giants'. You cannot land a spacecraft on Jupiter or Saturn because there is no hard surface. But both these planets have many moons, which are interesting to explore.

Titan—Saturn's Biggest Moon

Like Jupiter, Saturn has several moons. One of these, called Titan, is so large it is almost like a tiny planet. It is bigger than the smallest planet, Mercury, and it has an **atmosphere**. You cannot see the surface of Titan from space because the tiny particles and thick clouds that hang in its atmosphere hide everything from view. It just looks like a big orange ball. Huygens, a robot spacecraft, landed on Titan in 2005 and took photos of the ground showing that there are rivers and oceans. These don't contain water but are made of liquid **methane**. On Earth methane is a gas, but Titan is so cold that it turns to liquid there. Water also freezes and the surface is covered in 'boulders' of water ice covered in reddish chemicals, similar to those found in the atmosphere. These are oily chemicals known as hydrocarbons. The surface of Titan looks very strange.

How Do You Weigh a Planet?

To weigh other planets, we observe how hard their gravity pulls on their moons or on spacecraft we send to visit them. The heavier the planet, the greater its gravitational pull. The planet's weight determines how fast a moon will move in its **orbit**. Astronomers measure how far the moon is from the planet and how long it takes to complete an orbit. That allows them to work out how much a planet weighs.

The Moons of Jupiter

Jupiter has dozens of tiny moons, but there are four very large ones. One of these is called Io, which has volcanoes erupting on its surface. The others are Europa, Callisto and Ganymede. They are rocky, and covered with ice. Jupiter and its moons have been explored by robot space probes. No astronauts have gone to Jupiter because it is so far away. Even if we could get there it would be very dangerous. The space around Jupiter is filled with radiation energy, which could kill astronauts.

Jupiter

Jupiter is the largest planet in the **solar system**. It weighs more than two and a half times the weight of all the other planets put together—and nearly 320 times the weight of the Earth! Huge bands of clouds across its surface give it a striped look. You can see an enormous storm, like a cyclone, on its surface. It's called the Great Red Spot. This storm is larger than the planet Earth and may have been raging for as long as 400 years.

The Rings of Saturn

Saturn is almost as large as Jupiter. It has huge, beautiful rings made of particles of ice and rock around it. Nobody knows how these rings formed. Some scientists think that a moon of Saturn broke apart, scattering the ice and rock. Others think the particles in the rings fell into Saturn's gravity from elsewhere in space. The rings could have been formed when many large rocks collided over time, gradually pounding each other into small fragments. Smaller rings have been found around other planets such as Jupiter and Uranus, but they are not as spectacular as Saturn's.

Asteroids and Minor Planets

As well as Earth, there are seven other planets in orbit around the Sun. These are Mercury (closest to the Sun), Venus, Earth, Mars, Jupiter, Saturn, Uranus and Neptune. This is what is known as the solar system. Each of these planets has been visited by at least one spacecraft. The last planet to be explored was Neptune when a spacecraft called *Voyager 2* flew past it in 1989. But there are other objects that are smaller than planets that orbit the Sun. They are also being explored by spacecraft.

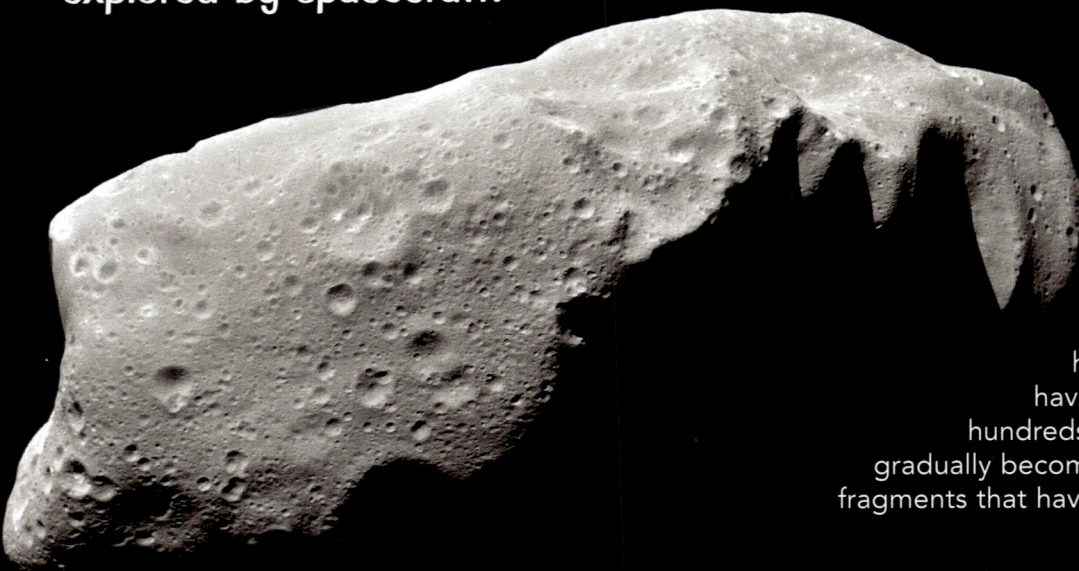

Outer Space Rocks

Asteroids are like huge rocks with strange, irregular shapes. There are thousands of asteroids orbiting the Sun between Mars and Jupiter in what is known as the 'asteroid belt'. Some asteroids also travel close to the Earth. Most asteroids have **craters**, like shallow holes, in their sides where other objects have crashed into them. Some are made up of hundreds of smaller rocks that have collided and gradually become stuck together over time. Others are fragments that have broken off larger objects.

Collecting from Comets

Comets are huge balls of ice and rock that look a lot like icy asteroids. Their surfaces are also covered in dust and rock. When comets get too close to the Sun the ice begins to melt. A huge 'tail' of water vapour and rock fragments flies out into space, which can be so large that it is visible from Earth with a telescope. It's dangerous to fly a spacecraft close to a comet when it is throwing off ice and particles. But some spacecraft have gone close to collect samples of the dust and rock. In 2006, the *Stardust* mission returned to Earth after nearly six years in space with a capsule of comet dust it had collected.

Exploring Asteroids

Spacecraft flying close to asteroids have discovered that they are covered in dust and rocks. A spacecraft called *Dawn* was launched in 2007 and will arrive at Vesta, a very large asteroid, in 2011. It will then go on to explore the asteroid Ceres in 2015. Vesta and Ceres are so large that they are also sometimes considered to be **minor planets**.

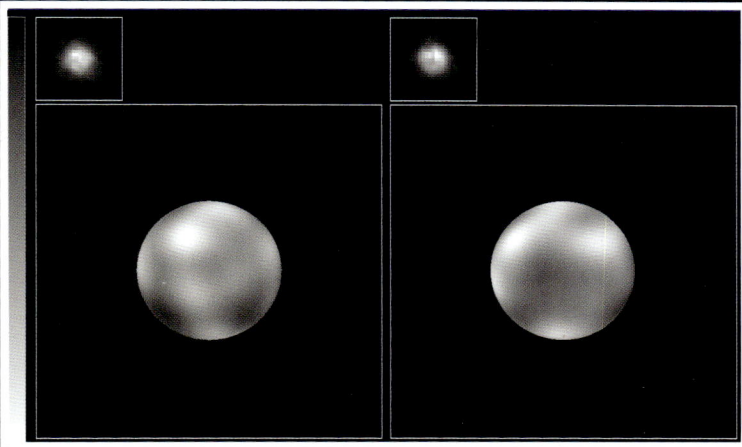

Minor Planets

In 1930 scientists discovered an object beyond the orbit of Neptune and called it Pluto. They used to think it was a planet, but now they know it is really too small to be a planet. It is called a 'minor planet' or 'dwarf planet'. Eris is another minor planet that is slightly larger than Pluto. A spacecraft called *New Horizons* is currently travelling towards Pluto. It will provide the first close look at this world in July 2015. Then it will travel further into space, to explore other minor planets. Pluto is so far away that even the most powerful telescopes cannot see it very well.

Asteroid Atmospheres

Asteroids and minor planets are relatively small objects. Because of this, they do not have much gravity. This weak gravity cannot hold much gas close to the surface so asteroids do not have atmospheres, and minor planets like Pluto have very thin ones.

Sleeping in Space

It's easy to sleep in space because you don't need a bed or a mattress. You can just float. It's comfortable but most astronauts use sleeping bags. Of course they have to attach themself to something so they don't float around and bump into things. They sometimes have to wear an eye mask to block out the light, because there's usually a light on inside the space station.

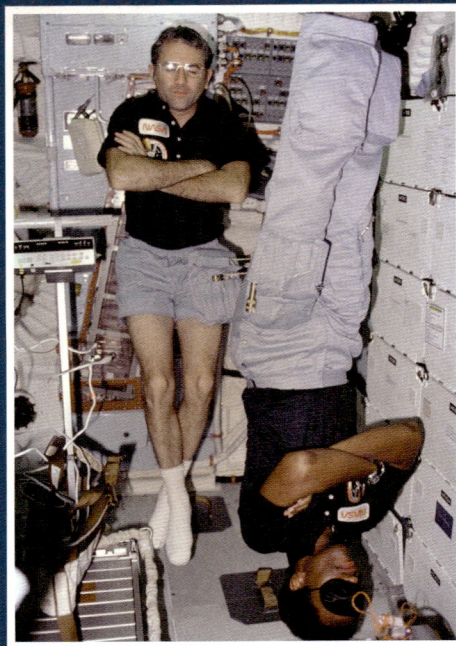

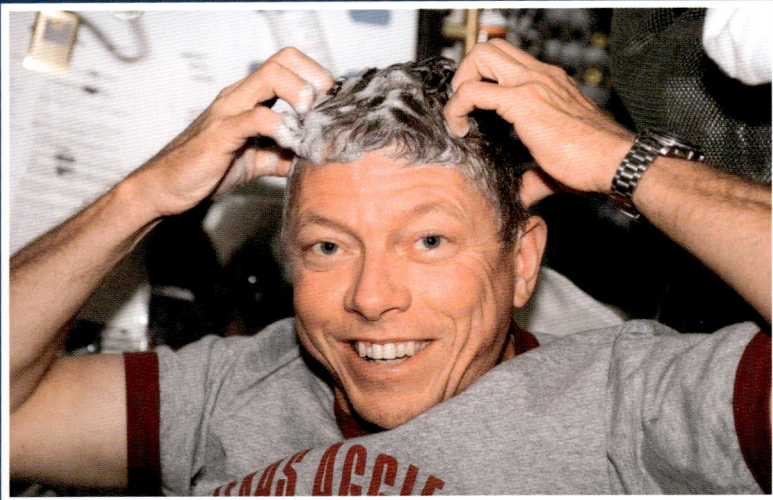

Washing

There isn't much water on board a spacecraft. You cannot take a bath, but some space stations do have showers. You unfold a plastic curtain to make a shower stall then you squirt yourself with water from a water gun. Because there is no gravity the water doesn't fall like a normal shower or run down the drain. You have to use a vacuum hose to collect all the water. You can wash your hair with special foamy soap that doesn't float away because it's sticky. If you have long hair, you may need to place a plastic bag over the top of your head to keep the water and soap in place.

The Toilet

Going to the toilet in space is difficult. There is no gravity to pull the waste away. A special toilet uses air currents to help move the waste away. It is stored in canisters, and sometimes dumped outside the spacecraft. NASA spent US$23.4 million to build a toilet for the space shuttle!

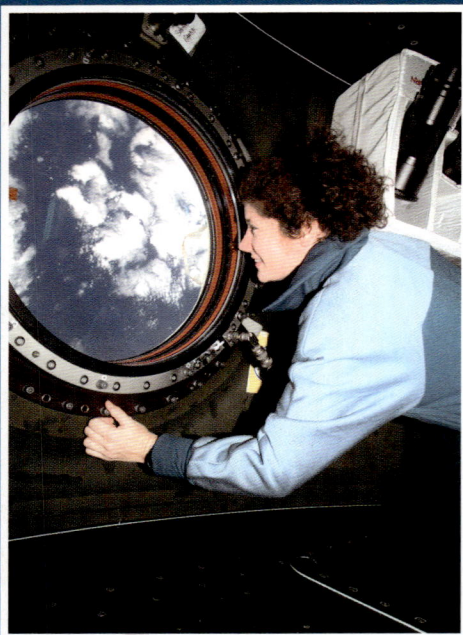

Watching the World Go By

The astronauts can do a lot of the same things that they do down here on Earth to have fun. They can watch movies on their computer screens or read books. A favourite pastime is simply looking out the window at the Earth. The view changes all the time, and they can see many different countries. There's also a spectacular sunrise or sunset every 45 minutes.

Living in Space

Things that seem very simple on Earth, like eating, sleeping or going to the toilet, can be very difficult for astronauts in space. Years of space missions have produced some interesting ways to solve these problems.

Space Food

Eating food in space is different from eating food on Earth. For a start you cannot use an ordinary plate, because the food would just float off. Food comes in cans or containers that have covers to keep the food inside. These containers have to be held in place on a meal tray. Some food is dehydrated so you have to add water before it can be eaten – just like instant noodles here on Earth. If you like pepper and salt it comes in a liquid form because small particles could float into your eyes or get stuck in equipment.

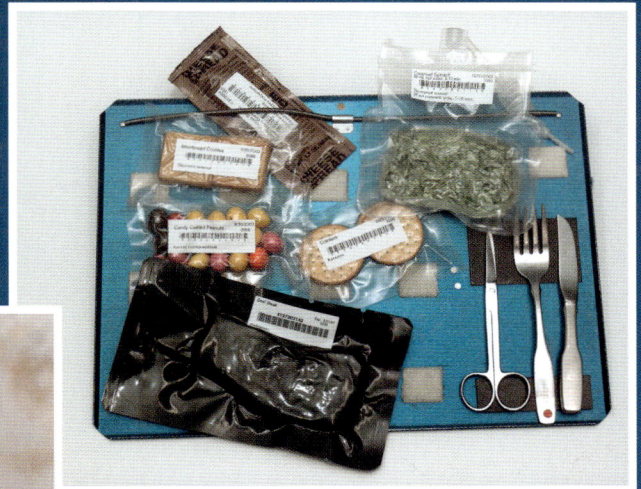

Floating Liquid

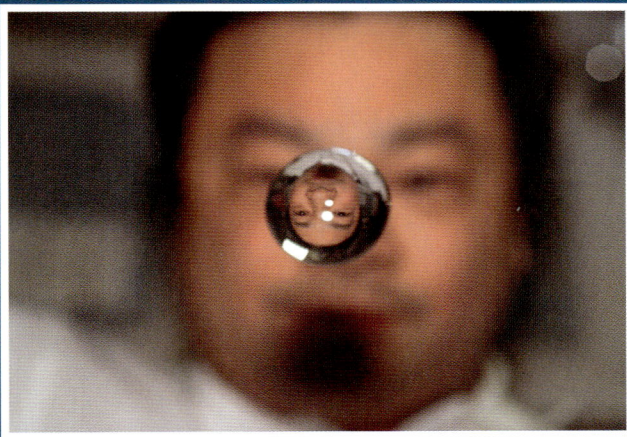

You cannot use a cup or glass in space because the drink would float out so drinks are kept in containers with special straws. These straws have clips that keep them closed when they are not being used. Sometimes the astronauts have fun with their drinks. If you squeeze some liquid out of its container, it will float in front of you like a ball. The more liquid you squeeze out, the bigger the ball becomes. If you want to drink it, you can just put your mouth around it and swallow or, if it's large enough, you can stick a straw into the middle of the ball and drink it.

What to Wear in Space

Astronauts wear special clothing in space—different types of clothes at different times. They don't always need heavy spacesuits. Sometimes they wear clothes that are just like normal clothes on Earth.

Launch and Landing Suits

Astronauts must wear special spacesuits during the launch, and when they are returning to Earth. These are the most dangerous times in any space mission. You have to wear a helmet on your head and gloves on your hands. Your whole body is covered in a thick suit. It is not very comfortable but it will keep you alive if the air leaks out of the spacecraft. Once you reach space, you can take this suit off and wear something more comfortable.

No Laundry to Do

Most of the clothes astronauts wear in space, apart from their spacesuits, don't get washed. When they are worn out they are simply put in with other disposable garbage and sent out with a supply vehicle on its return to Earth. All this rubbish is burnt up as it enters Earth's atmosphere — so there's no need for washing machines in space!

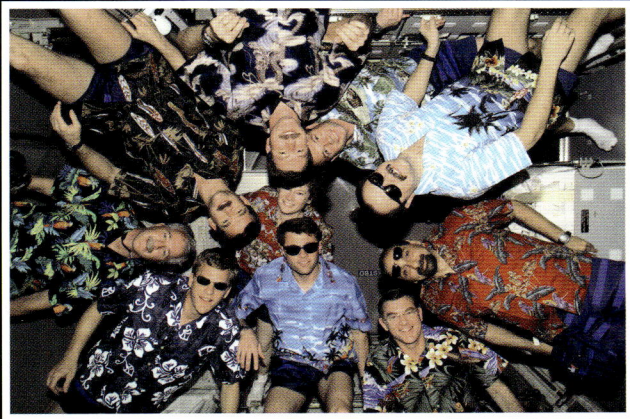

In Orbit

Astronauts in orbit wear fairly normal clothes. Sometimes they wear T-shirts and shorts. They also have coveralls, which are one-piece uniforms. Sometimes astronauts might brighten things up by wearing wild clothes, like Hawaiian shirts.

Sun Protection

The spacesuit worn on spacewalks is covered in white protective cloth that reflects the sun's rays to keep you cooler. You also have a visor on your helmet. If the sunlight is too strong, you can pull the visor down. It's like wearing sunglasses on the outside of your spacesuit, but it covers the whole of your face. To other people the visor looks like a mirror. They cannot see your face when the visor is down.

Walking in Space

To walk in space, you need a very heavy type of spacesuit. This has to be larger and stronger than the suits worn for launch and re-entry to protect you from exposure to space. You have a helmet and a backpack that gives you oxygen. You also wear a tight cap on your head, underneath the spacesuit. This holds earphones and microphones so you can talk to the other astronauts by radio. The spacesuit is so big that it's difficult to move in. Sometimes it feels like you are floating inside the suit! Your feet don't touch the base of the boots, and your head doesn't touch the top of the helmet. If you are doing a lot of work in space, you get tired quickly because the suit isn't very flexible. It's hard to move your fingers in the gloves, and your arms and legs don't bend very well.

Dangers of Space

Most people who travel in space return to Earth safely, but space is a very dangerous place for humans. Not only have astronauts died on space missions, but there are also dangers for people who stay in space for a long time.

Robot Space Probes

More unmanned, robot space missions have failed than missions with astronauts on board. But people don't worry about them so much, because nobody is killed on these missions. This is one reason why many space missions are flown without astronauts on board.

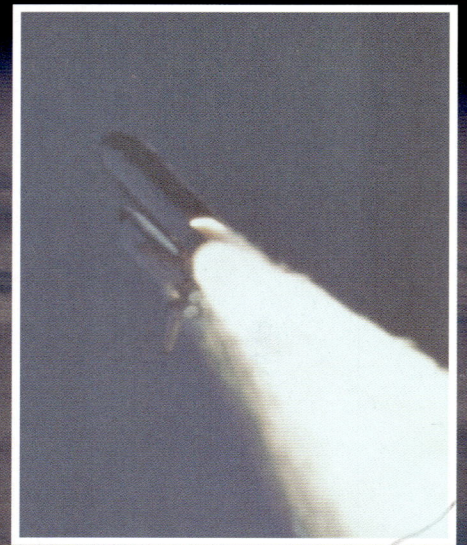

The Lost Shuttles

Two space shuttles have been lost during space missions. In 1986, *Challenger* broke up shortly after launch and *Columbia* was destroyed during re-entry in 2003. Fourteen astronauts were killed in these two accidents. Even though the space shuttle will continue to be used for a few more years, it is not regarded as a very safe vehicle to fly in space.

MIR Space Station

The Russian Mir Space Station, which is no longer in orbit, had many problems with its life support systems. Sometimes the oxygen generators broke down. Once, a major fire erupted inside the space station! Mir was eventually abandoned because it was too old to work safely.

Life Support Problems

Space travellers sometimes have serious problems with the life support systems on their spacecraft. In 1970, the *Apollo 13* mission was flying to the Moon. Suddenly, an oxygen tank on board the spacecraft exploded and it lost its air supply and electricity. Fortunately, the lunar module was docked to the main Apollo spacecraft. This module was supposed to carry the astronauts to the surface of the Moon, but now it became their lifeboat. They used air and power from the lunar module to stay alive while the spacecraft returned to Earth. They were lucky to get home.

Wobbly Space Legs

Living in space for a long time can lead to health problems. Muscles grow weak when the body is exposed to weightlessness for several months. This is because the muscles do not need to fight against gravity. Bones also grow weaker, and the body loses some of its resistance to disease. It is essential for astronauts to exercise regularly.

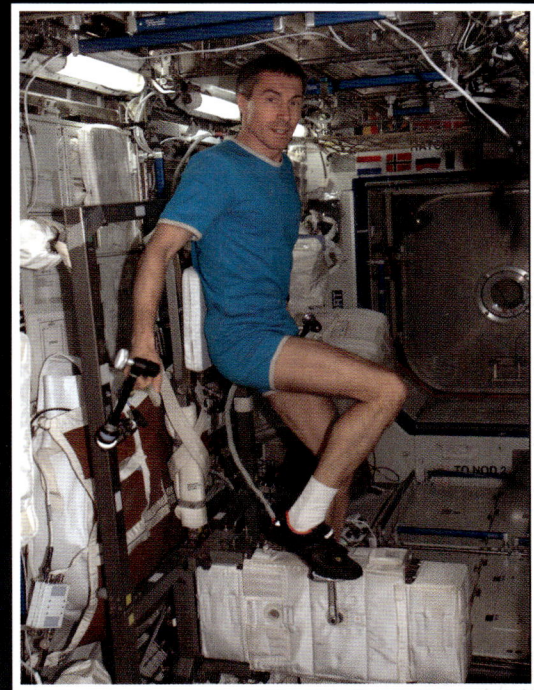

1967: Year of Tragedies

The first space traveller to die during a space mission was a Russian **cosmonaut**, Vladimir Komarov. In 1967 he was piloting the first manned mission of Soyuz, a new spacecraft. After he reached orbit, the spacecraft did not work properly and he was ordered to return to Earth. But the parachute on the *Soyuz* return capsule did not open fully, and he crashed into the ground. Also in 1967 three Apollo astronauts died, in a ground test before the first manned Apollo mission. They were Gus Grissom, Ed White and Roger Chaffee. Faulty electrical wiring caused a fire inside their spacecraft.

Search for Life

Planet Earth is the only place we have ever found life. Most planets and large bodies in space seem to have no living things at all. It's as if the Earth is an island of life in a vast sea of lifeless space. Conditions on most planets we know about are so dangerous and so extreme that it would be difficult for most forms of life as we know them to survive there. However, we also know that life can survive on Earth even in places where it is very cold, or very hot, or where there is no oxygen or sunlight. So it might still be possible for life to exist on these other worlds.

Is Anybody Out There?

Nobody has found life in space yet, but this doesn't mean that there is no life on other planets. There could be simple forms of life on other planets in our own solar system. There could be more complex life forms, like intelligent creatures, elsewhere in the universe. Scientists are searching for life elsewhere in space. In a few years, they could succeed.

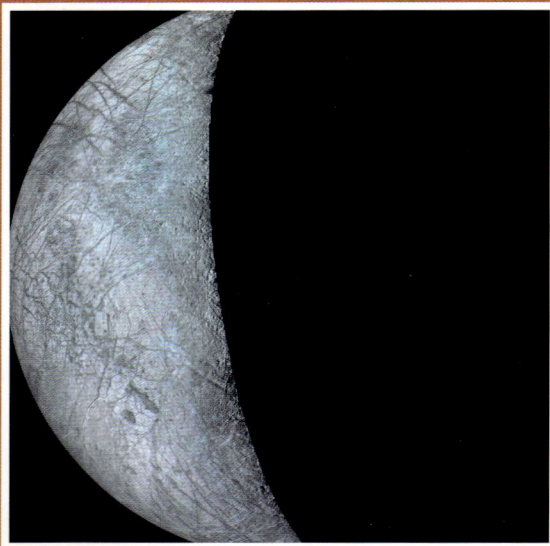

Searching for Intelligent Life

Scientists expect that the first life forms that are found in space will probably be simple, microscopic creatures. These will probably be discovered on a planet somewhere in our own **solar system**. Jupiter and Saturn, for example, have moons that are covered in thick layers of water ice. Scientists think that there are oceans of liquid water beneath the layers of ice that could be filled with life. Eventually, spacecraft could be sent to explore them and search for life. But it would be more interesting to find more complex forms of life. There could be alien life forms that are as smart as humans.

How Will We Find Life Out There?

Scientists are using huge antennas called radio telescopes to try to pick up radio transmissions from aliens in space. They might be sending messages to tell other planets that they exist just as scientists on Earth send radio signals out into space, to let other planets know that we exist! So far, nobody has heard a transmission from another planet. But the universe is very big. It could take many years, or even centuries, to hear from intelligent life somewhere in space.

Laser Beams

Some astronomers think that extraterrestrials could try to communicate with us using laser beams. They use special telescopes to try to detect laser beams from deep space. So far nobody has detected a signal.

Is There Life on Mars?

While nobody thinks we will find little green men on Mars there could be life hidden deep below the surface. On Earth, small, microscopic creatures can be found deep underground, feeding on chemicals in rocks. Similar life forms could also exist on Mars. In the future, rocks drilled from far below the Martian surface could be flown back to Earth by a robot spacecraft. Scientists might then discover life inside them.

Space Tourism

Most people who have flown in space are professional astronauts. They have to be selected by their governments to fly in space, then they have to train for many years. But it will soon be easier for people to become space travellers. Just like passengers on an aircraft you could pay to fly on board a spaceship. This is called space tourism.

The New Spacecraft

Companies around the world are developing new spacecraft, just to carry tourists into space. They will be much cheaper to fly on than *Soyuz*, but they will still be expensive. You could pay more than $200 000 to become a space tourist. These new spacecraft look like aircraft, and can only stay in space for a few minutes. They don't go into orbit around the Earth, and they cannot reach the International Space Station. But they do allow you to see the Earth from space, and feel what it's like to be weightless for a short time.

Cheaper Flights

In the future, the cost of space tourism will gradually get cheaper. It could cost only a few thousand dollars to fly in space. Many people will become space tourists. You could be one of them.

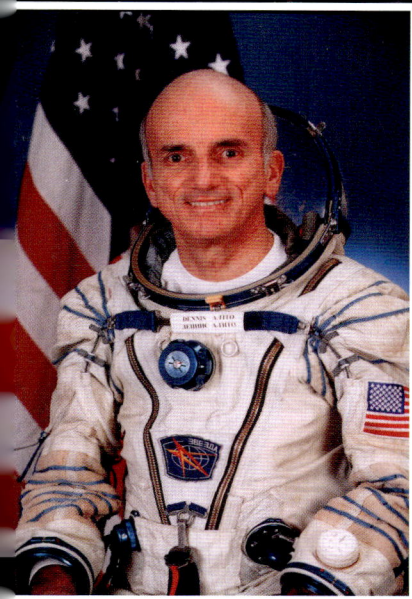

The First Tourists

The first space tourist flew in 2001. He was an American businessman named Dennis Tito. He paid Russia US$20 million for a seat on board a Russian Soyuz spaceship, and flew to the International Space Station. Dennis Tito did not pilot the spaceship, but he had to spend months learning how to use its controls. Other people have also paid to fly on Soyuz spaceships but the number is small because very few people can afford the cost of such a trip.

Being a Space Tourist

Before you fly in space, you spend a few days learning about your spacecraft and what to do in the event of an emergency. When you go on your flight you will wear a special uniform with a helmet that will give you oxygen. In the spacecraft you and the few other passengers will have a huge seat and a large seat belt. As the spaceship takes off like an aircraft from a runway, you will hear jet engines roar as you climb into the sky. Then, the jet engines switch off and a with a thunderous wail the rocket motors are switched on. You are pushed back into your seat by the force. The sky turns dark blue, then black. You are in space so you will feel lighter, and you can unbuckle yourself and float around the cabin. The Earth outside looks beautiful. Before long the pilot will tell you to strap yourself back in for the return to Earth. You feel heavier as you re-enter the **atmosphere**. Then the spaceship glides towards a runway. You have made it home.

Big Space Programs

Only three nations, Russia, America and China, have built rockets and spacecraft that can launch people into space. Russia and America launched their first space travellers in 1961. The first Chinese astronaut made the journey into space in 2003. America is the only nation to have sent astronauts to the Moon, and it has also sent robot probes to every planet in the **solar system**. Russia has sent robot probes to the Moon, Mars and Venus. For many years, Russia was the leading operator of space stations. Russia and America now work together on the International Space Station.

International Projects

Some satellites or missions to the planets are international. Several different nations contribute parts and money to the mission. This means that one nation doesn't have to pay for the entire cost of the spacecraft. Everyone gets to see the results. It's like getting a regular space mission at a cheaper price.

Smaller Programs

Some teams of university students have built their own small satellites. Being small, they are easy to send up into space either by small rockets or with a space shuttle. This allows a country without a lot of money to take their place in the space age. Some of these satellites are not very large. They are often about the same size as a microwave oven, or even smaller. They can still take measurements in space and transmit their findings back to Earth.

Guest Astronauts

Dozens of other countries have launched people into space as guest astronauts on Russian or American spacecraft. They train to operate the spacecraft and then join a crew. Some space missions can have people from three different countries on board. Usually, the countries sending the guest astronauts have to pay the costs for their own astronauts.

Around the World

Many nations operate satellites for communication or observing the weather. But most do not have the resources to run huge space programs capable of launching satellites into orbit or sending people into space.

Advanced Programs

Some nations that cannot launch their own astronauts still have advanced space programs. Several nations in Europe have formed the European Space Agency. Together, they launch their own rockets, send probes to the planets and, in the future, they could build a spacecraft for launching astronauts. Japan also builds its own rockets and satellites, including robot spacecraft that have explored the Moon. India has developed rockets and satellites and could send missions into deep space in the future. Other nations such as Canada operate advanced satellites but do not have their own launch rockets.

Back
to the Moon

Apollo 17, the last manned mission to the Moon, landed in December 1972. It's been a long time since astronauts have explored the Moon. But in the future, they will return there. They will do more work on the Moon than the Apollo astronauts, and stay for much longer.

Landing on the Moon, Again

The lunar module used in the Apollo missions was tiny and cramped. It could barely fit two astronauts on board. The next lunar module will probably be able to take four or more astronauts to the surface of the Moon. Its design is currently being developed and will take several years to complete. Astronauts could return to the Moon again some time before 2020. The first astronauts to go back there will probably all be from America. They will land in a region that was not explored by the Apollo astronauts and will stay on the Lunar surface for a week, or longer.

A Base on the Moon

Eventually, a base for astronauts could be built on the Moon. The first base would be made from a few crew modules, similar to those used on the space station. But a larger base would probably be built underground to protect it from heat, cold and the dangerous radiation from space. Astronauts could live there for a few months at a time. The interior of the Moon Base would look like a space station. It would take many missions to carry enough parts to build the base, but the astronauts would also use rocks and soil from the Moon itself to help build it.

Many Nations

The Apollo astronauts who went to the Moon were all Americans, but the crew of a large Moon Base would probably come from many different countries. It will cost more to build a base on the Moon than a space station, because it is much further away. Spreading the cost over several nations would make it affordable.

Robots on the Moon

The astronauts will take robots with them to the surface of the Moon. Some of these will keep working after the astronauts have gone home. The robots could move around and explore further. Some of them will have wheels. The next lunar rover could work like a robot when there are no astronauts on board, and drive itself.

Telescopes on the Moon

The Moon is a good place to use telescopes because there is no air to block the light and there are never clouds in the sky. It is also a long way from the Earth's magnetic field. This means that some instruments, such as radio telescopes, can work better. Astronauts could carry the parts for the telescopes and assemble them after landing. When the astronauts leave the Moon, the telescopes will keep working by remote control. Astronomers on Earth will be able to look at the pictures sent back on a radio link.

Asian Moon Probes

In 2007, China launched its first robot probe to the Moon. Japan and India are also exploring the Moon with space probes. Eventually China could send astronauts to the Moon.

Humans on Mars

Nobody expects to launch astronauts to Mars in the near future. It is a long way to travel, and the conditions in deep space are more dangerous than in Earth's orbit. But eventually, astronauts will explore the planet. It is not expected that the first astronauts will be launched to Mars before 2020.

Fast Travel

Our current rockets would take longer than a year to send astronauts to Mars. If something went wrong with the spacecraft, it would be almost impossible to rescue the astronauts. The spacecraft and its life support system would need to be more reliable than our current spacecraft. One way to make the mission safer would be to fly to Mars very fast so an advanced type of propulsion would be needed. This could use nuclear energy but it has not been tested well enough to build a Mars spacecraft yet.

Big Spacecraft

A spacecraft launched to Mars with humans aboard would need to be very big. It would probably be larger than a space station because it would need room for supplies, the crew, and a large propulsion system with fuel. The landing module that carries astronauts to the surface of Mars would need to be strong. It would need a heatshield to protect it from the atmosphere of Mars, otherwise, it would be destroyed as it descended. The module would need to be protected against the fine dust on the planet, which could enter into mechanical parts and jam them.

Dangerous Planet

Some scientists think that the dust on the surface of Mars could be poisonous. Astronauts could become sick if they are exposed to the dust, or breathe it in. It would be essential to clean all the dust off an astronaut's spacesuit before they return to their spacecraft. Otherwise, they would carry dust inside the spacecraft, where it could be inhaled. It would also be important to pick a safe landing site where there are no dangerous rocks. Some rocks are so large on Mars that a spacecraft could tip over as it touches down.

Keeping Mars Clean

Some scientists don't want any astronauts to go to Mars until it has been thoroughly tested for life. They are worried that the astronauts will bring germs from Earth with them that could contaminate Mars and interfere with any life that might be on the planet.

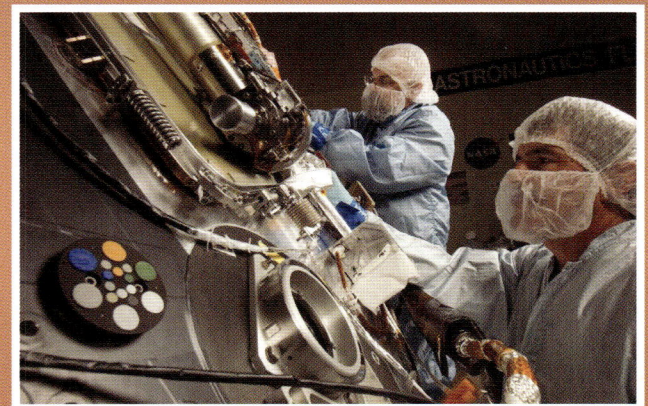

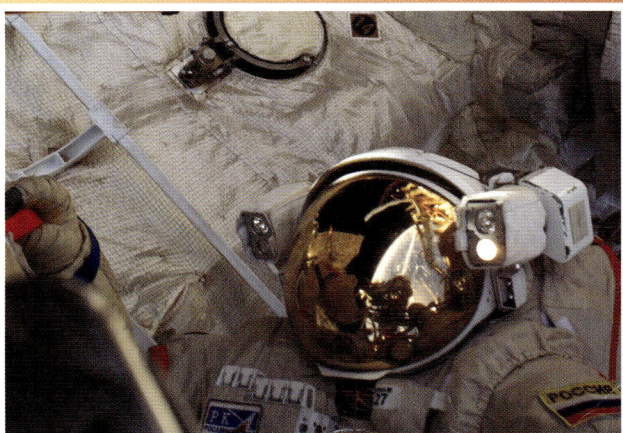

Coming Home

When they return from Mars, astronauts will need to be checked carefully by doctors. A long space mission can affect your health. The doctors will also want to know if the astronauts have been infected with any germs from Mars, which could be dangerous to life on Earth. The astronauts will be kept in isolation until their health is checked.

The Golden Record

Two Voyager spacecraft were launched into space in 1977. Their paths have carried them beyond the **solar system** towards the stars. Each spacecraft carries a greeting for any aliens that might find them. It is recorded on a gold-plated disk, roughly 30 centimetres across. The disk contains sounds and pictures that show the richness of life on Earth. There are sounds from nature, music from different cultures and eras, and spoken greetings in 55 different languages.

Going to Other Stars

At the moment, nobody knows how to send astronauts to other stars. They are so far away that it would take too long to go there. Even the advanced propulsion systems in development would take centuries to get there! But some scientists think it could be possible to 'warp' space. This means that space, time, gravity and other physical properties could be controlled with special equipment. Nobody really knows how to do this yet, but they are working on it. This could make it possible to travel huge distances without a rocket, in a short space of time. Maybe you will be involved in finding out how to get to the stars!

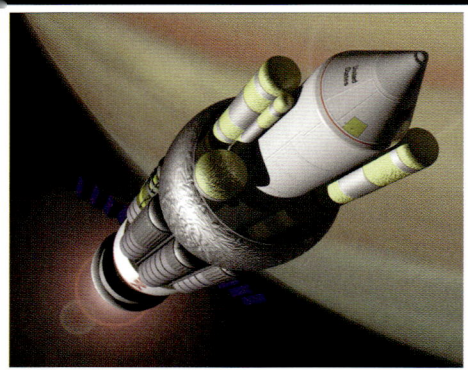

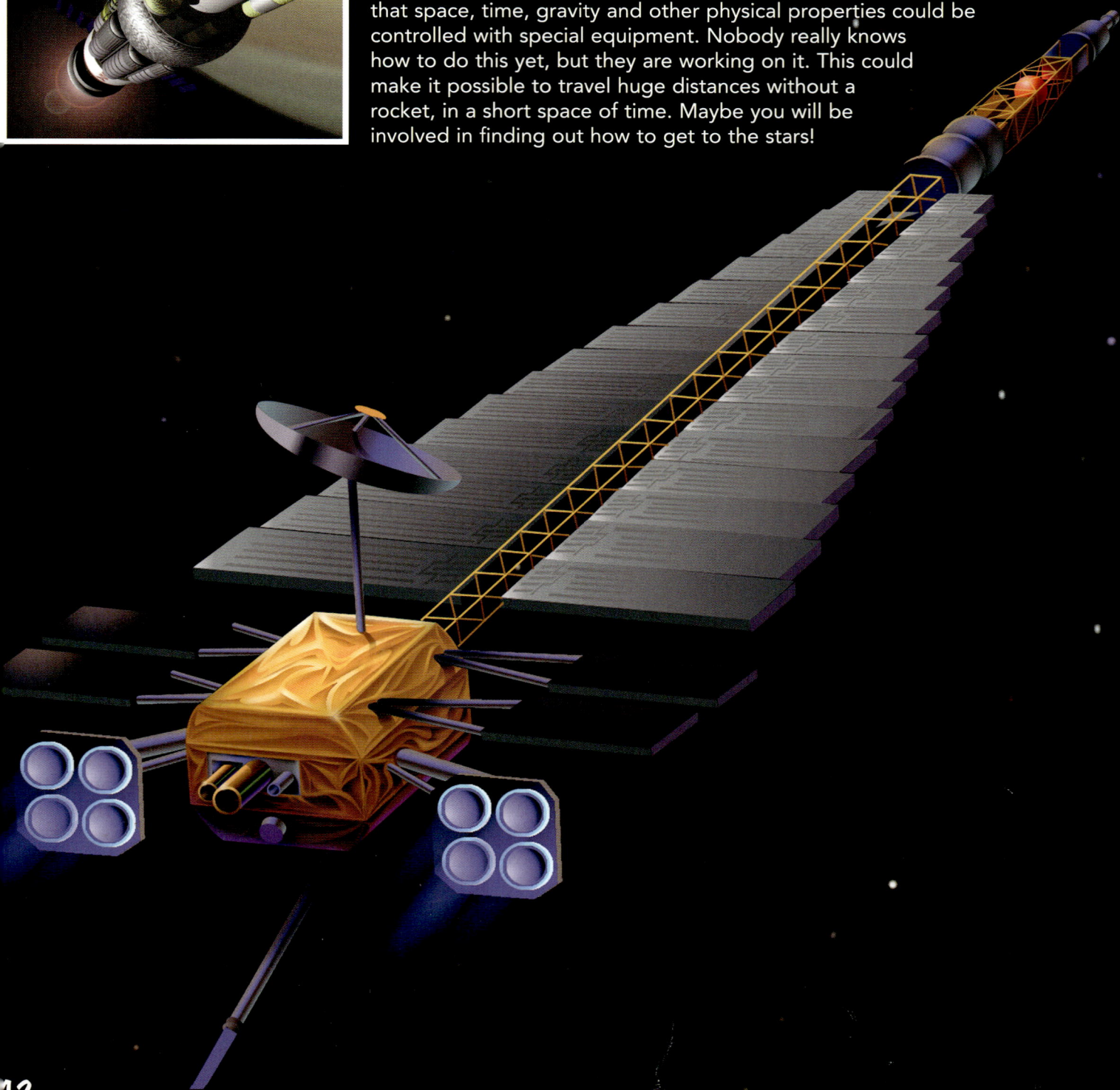

What Next?

Space travel has existed for more than 50 years, but we have only taken our first few steps into the universe. In the future, astronauts will go where no one has ever gone before.

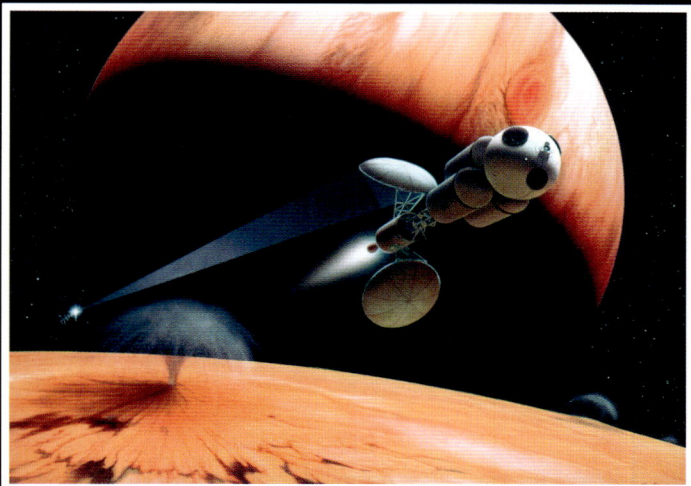

How Far Can We Go?

No astronauts have gone beyond our solar system and it is very difficult to send robot space probes there. The rockets we have now are not fast enough. In the future this will change, as new forms of propulsion are developed that can make spaceships travel faster. Some of these are already being tested. They do not use rocket fuel. Some of them use electricity, magnetism and other forces to generate thrust. One day it is expected that astronauts could go to the edge of the solar system and explore Neptune, Pluto and Eris, then return to Earth with samples they have collected.

Bed and Breakfast in Space

Some businesspeople are already drawing up plans for hotels in space. You could ride a spacecraft into orbit and dock with the space hotel. This would be like a huge space station, but the people inside it would go there for holidays. You could play sports in weightlessness, and look back at the Earth. Some tourist spaceships could even take you all the way to the Moon – and beyond.

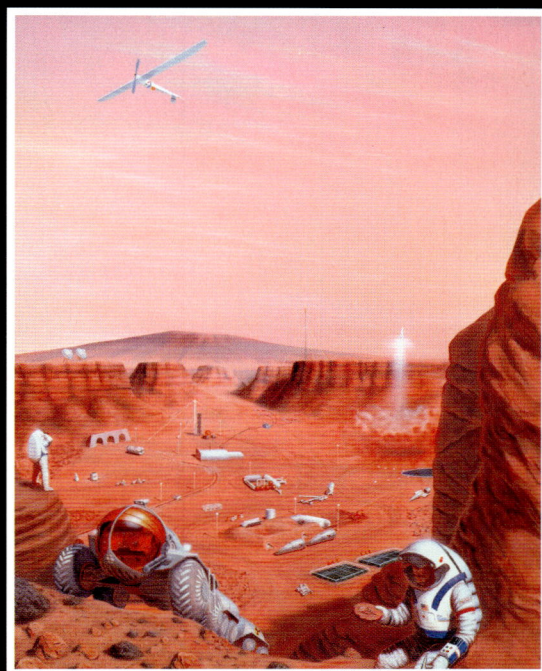

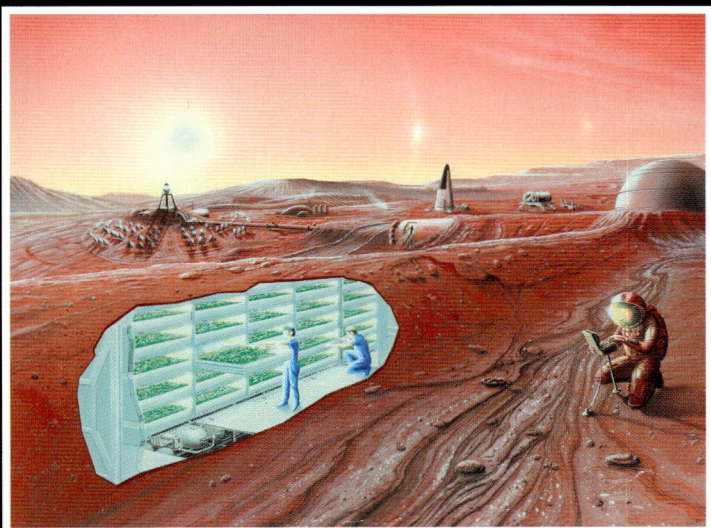

Would You Move to the Moon?

Some nations could build large bases on the Moon and Mars. Hundreds of people could live there for years. Some bases would be built underground and plants could be grown to produce oxygen to breathe and food to eat. It is not possible to do this now, but new machinery yet to be invented could allow us to do it in the future.

Activities

The answers to all these questions are all in the book. To help you, the page numbers you should look at are given for each question. Check your answers on page 47.

1. What is the largest moon of Saturn? (pages 22–23)

2. How many astronauts have walked on the Moon? (pages 18–19)

3. Who was the first space tourist? (pages 34–35)

4. What is the name of the spacecraft heading towards Pluto? (pages 24–25)

5. What is the maximum size of a space shuttle crew? (pages 12–13)

6. In what year did the two Mars Exploration Rovers land on Mars? (pages 20–21)

7. Which Apollo mission suffered from an oxygen tank explosion? (pages 30–31)

8. Which spacecraft was the first to reach the planet Neptune? (pages 24–25)

9. What are the oceans on Titan made of? (pages 22–23)

10. What is a satellite that monitors navy ships and tanks called? (pages 10–11)

11. What was the name of the world's first satellite? (pages 8–9)

12. Is Eris a planet? (pages 24–25)

13. How many planets are known to have life? (pages 32–33)

14. Which is more powerful: Solid or liquid rocket fuel? (pages 6–7)

15. What did Alexei Leonov do? (pages 8–9)

16. Who were the three astronauts on Apollo 11? (pages 16–17)

17. Which two spacecraft carry astronauts home from the International Space Station (pages 14–15)

18. Which nation was the third to develop its own astronaut-carrying spacecraft? (pages 36–37)

19. Which are the two most dangerous phases of a space mission? (pages 28–29)

20. Which was the last Apollo mission to land on the Moon? (pages 18–19)

Glossary (what words mean)

astronaut A person who travels in space. This is the usual term for people on board an American spacecraft.

atmosphere A layer of gas covering the surface of a planet or other large object. Some planets are mostly made of their atmospheres.

cosmonaut A space traveller from Russia. Also used for space travellers from other countries who fly on Russian spacecraft.

crater A roughly circular hole caused by an object colliding into the surface of a Moon, planet or other object in space.

methane A hydrocarbon chemical, normally a gas on Earth. It becomes liquid at very low temperatures.

minor planet A type of object that is considered too small to be a planet, but too large to be an asteroid. Minor Planets are usually spheres, and some have thin atmospheres.

moon A large natural object orbiting a planet. Some planets have more than one. Others have none.

orbit The path of motion of a satellite around another large object, such as a planet.

rover A vehicle that drives around on the surface of the Moon or a planet. Some rovers can carry astronauts. Others are remote controlled.

solar system A sun and everything that revolves around it, including planets, asteroids and comets.

Want to Know More?

If you want to know more about space and astronauts, try the following websites and books.

Websites

NASA
www.nasa.gov
This is America's Space Agency. NASA operates the Space Shuttle and the International Space Station.

Jet Propulsion Laboratory
www.jpl.nasa.gov
The Jet Propulsion Laboratory is a branch of NASA that launches missions to the planets.

European Space Agency
www.esa.int
The European Space Agency builds satellites and space probes.

Space Adventures
www.spaceadventures.com
Want to be a tourist in space? Space Adventures has information.

Japan Space Exploration Agency
www.jaxa.jp
Japan's Space Agency has a good section in English on their Moon program.

GeoEye
www.geoeye.com
GeoEye operates satellites that photograph the Earth. Can you see your home town?

Weighing a Planet
http://spaceplace.nasa.gov/en/kids/phonedrmarc/2001_december.shtml

Answers to Activities

1. Titan
2. Twelve
3. Dennis Tito
4. *New Horizons*
5. Eight
6. 2004
7. *Apollo 13*
8. *Voyager 2*
9. Liquid methane
10. Spy satellite
11. *Sputnik*
12. No
13. Only the planet Earth
14. Liquid rocket fuel
15. First person to walk in space
16. Neil Armstrong, Buzz Aldrin, Michael Collins
17. Space shuttle, Soyuz
18. China
19. Launch and the return to Earth
20. *Apollo 17*

Index

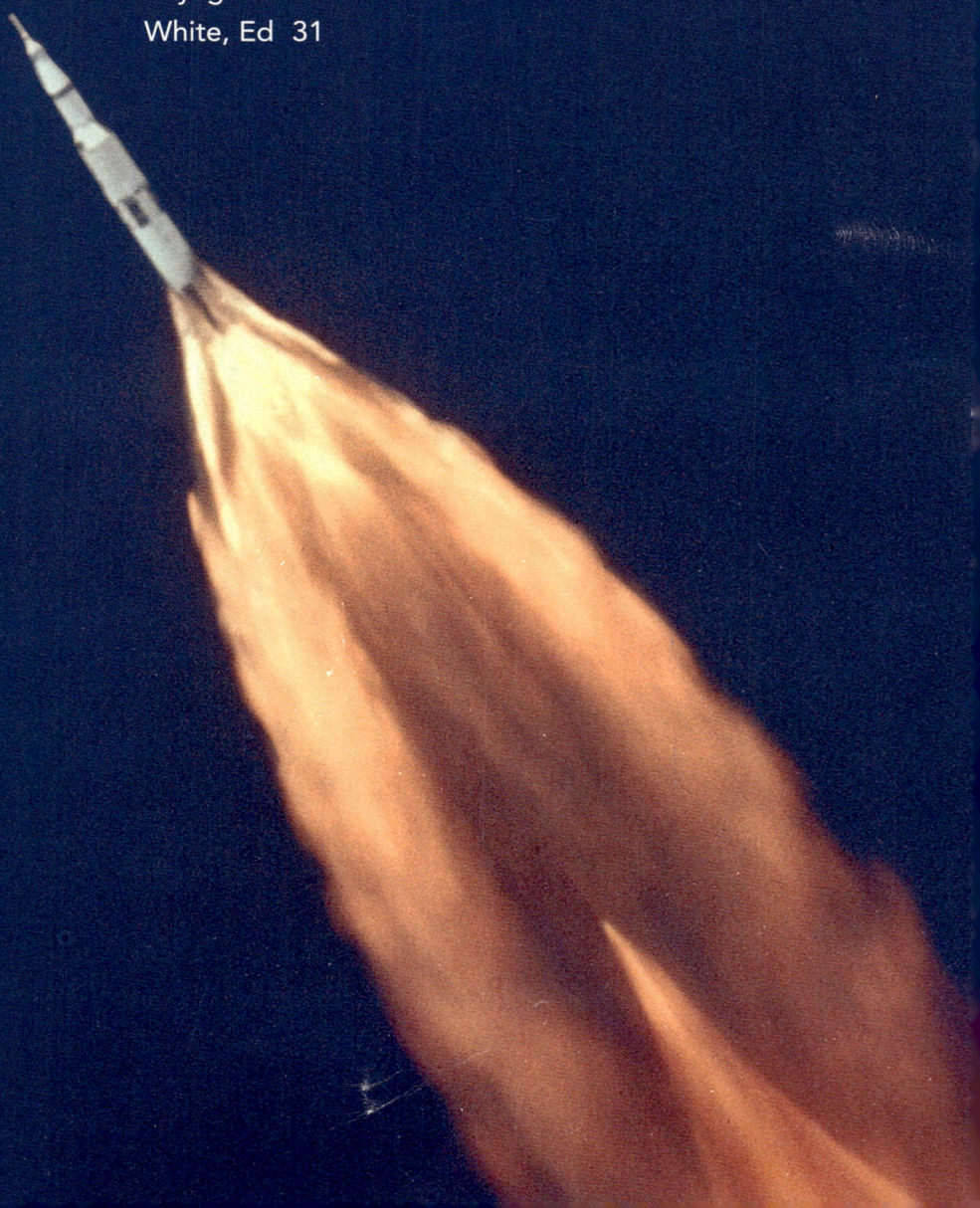